67 Need To Know Tips To Extend Your Car's Life

By: Jim Gaines

Jim Gaines

Text Copyright © 2014
WCM PUBLISHING All Rights Reserved

No right to redestribute, copy, amend or exploit materials. The information presented represents the view of the author at the date of publication and not the publisher, and by rate at which conditions change, the author reserves rights to update opinions based on new conditions. Neither author nor publisher assume any responsibility for errors or omissions. This book is in no way endorsed or distributed by any brand/company/site/etc. listed herin, are the sole opinions of the author, and should be treated as such.

67 Need To Know Tips To Extend Your Car's Life

Preface To Car Care Tips

Welcome to 67 Need To Know Tips To Extend Your Cars Life. I'm Jim, and the only thing I love more than driving my car is driving it forever.

Being in the auto industry, I see and hear about all of the stories from drivers all over the world (mostly the United States), and have found that most of the issues I hear can be solved with some easy preventative maintenance.

And that's why I wrote this book. I wanted to give you the EASY tips and things you can do BEFORE bad stuff happens, so that it will never happen in the first place. Wouldn't you like to know that by spending a little time and money here and there, you will save countless hours waiting for mechanics to figure out how much to charge you for your lack of car care, and countless money from the unnecessary repairs and maintenance that wouldn't happen in the first place if you followed these tips?

The money and time it takes to do what I have laid out in this book is pocket change compared to the hundreds and usually thousands of dollars you'd have to spend later getting your car fixed.

You might already know about or follow some of these tips and strategies. If so, great! What I want you to do is not look at this book as the "bible of car maintenance" or the end-all be-all book of mechanics. I want you to look at this as a series of tips, strategies, and notes (with some stories sprinkled in) that if followed, will help your car last forever, will save you money, and will make you feel more knowledgeable and prepared for car troubles that seem like nothing but could end up costing you a lot in the long run.

So, car owner... driver... auto friend... Take the time to read the 67 tips I've laid out in this book, research further anything you'd like to know more about (although most of these tips are easy enough to do right away without any further info needed), and start living a life with less car troubles.

Enjoy the book, the tips, and the auto knowledge.

1

New Cars, Keep Speed Low

Most cars have what is called a "break-in period".

This is usually the period in between 500 to 1,500 miles, but can vary depending on the type of car you have. It's best to check your owner's manual to see your car's specifications.

During this break-in period, it is recommended that you drive well below the common freeway speed limit. For me, I like to take things safe, so when breaking in a new car, I always stay below 55 MPH.

If you go too fast, and break 55 or 60 MPH during the break-in period of your new car, the piston rings may not settle into place the correct way. When this happens, your car can begin to leak oil, and as you might already know, this is horrible, especially for new cars.

Your oil starts to leak, your car starts to make funny noises, and before you know it, your engine overheats and completely busts. Now you have to buy a new car again.

Don't let that happen to you. Keep your speed to a minimum, and avoid highways (or at least the fast lane) for the first days and weeks of your new car.

Jim Gaines

2

No Long Idling

You may have heard the myth that you should "let your car warm up" by starting it and doing nothing for awhile. This puts your car in the idle state, and is called idling your car.

This is just a myth. You may think idling your car helps it, but it doesn't help at all - in fact it can hurt your car's engine greatly if idled for too long.

Idling doesn't send enough oil through the engine, and actually sends more fuel into your engine oil. Long story short, fuel in your oil screws up the oil's viscosity (which I'll explain in a later tip), and makes it difficult to lubricate your engine.

Also, the powertrain is not designed to idle for long periods of time, and the exhaust system is too cool for long periods of idling.

Now, if you are thinking, "well I do this all the time and have had no problems", good for you, but sometimes things like this don't happen on the first idle. In fact, this is more of a long-term effect. If you constantly "warm your car up" by idling every single day, it could hurt your car in the near future, leading to more problems, more repairs, and more money out of your pocket. The biggest problem is... So many people blame the car itself for not functioning a year later, and if they were to just follow these simple tips, it could be avoided all together.

If you really want to warm up your car, lightly hit the gas pedal, and increase your rev (without too much revving) slowly. If you are just warming up your car to defrost the windshield, invest in a squeegee, and take care of your visibility before you even start your car.

3

Avoid Heavy Loads

True story: My mother once went to a lumber yard, and loaded the entire back of her car with gravel. Although the car was okay during the drive home, once she parked the car, her back tires went completely flat, and because there was so much weight on the wheels, the rims began to bend, and she had to pay for 2 brand new tires and rims.

This could be even worse for you, so trust me, avoid heavy loads on your car.

Whether you are packing stuff on the top of car, in the trunk, or all throughout the car, make sure you understand how much weight your car can actually hold. Just because your car still rolls doesn't mean it's not going to hurt your car's health.

First, loading too much in or on your car is a safety hazard for both the people inside your car, and the people and cars around you.

Next, your suspension can completely give out, or worse, the frame and other important parts of your car can become weak, giving you an immediate bill to fix your car.

Besides direct hazards, if you load your car with too much crap frequently, you can expect many unnecessary repair bills in the near and far future.

Check your owner's manual, find out what the max load capacity is for your vehicle, and stick to it. I actually recommend staying well below the maximum, anything to keep my car running safe and steady for many years.

Jim Gaines

4

About Revving The Engine

Sure, you might love revving your engine in front of your friends, but is it all really worth it?

Especially at the start up of the engine, it is recommended that you never rev your engine, as it could add years of wear to the engine.

Revving the engine causes many things to happen, especially when your car engine is cold or just starting up.

It causes friction wear on cylinder walls, piston rings, and pistons, it causes excess gas to get into the exhaust, igniting in your catalytic converter, and even melts the ceramic emission filter, which is like stuffing a dozen tennis balls in your tailpipe.

Your gears will also still turn (even when the car is not in gear), making your gears wear down at a much faster pace.

So, if you do rev your engine, only do it once, and make sure your car has been running for awhile, and the engine is hot. But, not only does revving your engine make you look like a total douche, it also creates dozens of problems that can cause your car to diminish at a 10x quicker rate.

5

Don't Start AC Right Away

This is not only for your car's health, but is also for your own health, and I thought it best to include it in this book, as it is very important, and most people I talk to don't know about this.

Research has showed that the car dashboard AC units emit Benzene, which is a cancer causing toxin. Benzene poisons your bones, causes anemia, and reduces white blood cells. The more exposure to Benzene, the greater chance of Leukemia, cancer, miscarriage in women, and other diseases and cancer (this also affects your kidney and liver).

Now, this is nothing to worry about SOMETIMES, but other times it could be fatal. Let me explain...

A car parked indoors with the windows closed may contain about 500 mg of Benzene. That same car parked outdoors under the sun could have Benzene levels of 4,000 mg. If it's a really hot day, the Benzene levels skyrocket even further.

This research has shown that when you get into your car and keep the windows closed, you could be inhaling excessive amounts of Benzene and other toxins.

Most people believe that to get the hot air out, they should immediately crank up the AC, which can actually cause more toxins to enter your body, giving you a greater chance of the cancers and risks I mentioned before.

Here's what you SHOULD do: Once you get in your car, start it up, and roll the windows down. Then, turn on the fan/air in your car, WITHOUT AC, and give it a good 10 to 20 seconds to let the Benzene escape out of the windows of your car. Only then you can turn on the AC, as much of the Benzene will be gone.

Also, another tip: Your AC won't even work the right way when you first start your car, so it's already pointless to try and crank it up before you start moving.

Jim Gaines

6

Red Light Neutrality

This is what I like to call "Red Light Neutrality".

It's a trick I learned from my brother who is an auto technician (he picks up all sorts of fun tricks of the trade).

This can work while you are at red lights, waiting in line at a drive through, and other places you commonly stop your car, but keep it in D - drive.

Guess what? The longer you wait completely still in your car while in DRIVE, the more strain it puts on your engine.

Basically, when you put your car in DRIVE, you are telling it that you want to move forward. If you sit around too long, it puts so much unnecessary strain your engine it's unbelievable.

So, in order to fix this, every time you stop at a red light (especially the long ones), put your car in neutral, and you won't cause as much strain on your engine.

Same goes with drive-thrus. I bet at least once, you've seen somone's reverse lights come on once they stop, and it may have startled you. I know it's bugged me - I remember one time when I saw reverse lights come on in front of me, I laid on my horn for about 5 seconds thinking they were going to back into me.

But, I now know that they were just putting their car into neutral (or park) to take strain off of their engine. This will make your car last longer than usual, and will allow you to sit back and relax while other people have crazy problems with their cars.

7

Extreme Hot/Cold Speeds

This tip is going to be short and sweet, but nonetheless, it is VERY important!

I hear it every morning - My neighbor gets in his car, starts it up, and immediately speeds off. He goes from 0 to 30, then from 30 to 50 within seconds, and I laugh every time I hear it (he also takes his car to the mechanic much more often than I do).

Here's the lowdown: NEVER accelerate quickly in very hot or very cold weather, and NEVER drive at high speeds in very hot or very cold weather.

I'll explain more about what exactly could happen in situations like this later, but basically, doing these things can lead to more frequent repairs.

Remember earlier when I told you about the 'break-in' period of a car, and how you should keep your speeds to a minimum? This works sort of the same way. When it is extremely cold, your engine needs to work harder to become hot, so by accelerating quickly, you are straining your engine and many other parts under your hood.

It works the direct opposite when your car is too hot. If it's 90 degrees out, and you are pushing your car's engine to the max, well, it does just that - you are maxing out your engine, and making your car work harder to get your desired accelerated result.

So, next time you start to accelerate quickly in extreme weather conditions, think again. It may get you to your destination 10 seconds quicker, but those 10 seconds each day could lead to $1,000s in repair bills later. But of course, if you are rich, and can afford to crap up your car, go for it, no one is stopping you.

8

Burning Rubber

Another pretty self-explanatory tip, but again, I hear tires burning away every day, so obviously people are either rich enough to pay for new tires and suspension parts, or they are just stupid. Which one are you? ;)

Okay, really though, why in the hell would you burn rubber? I'm talking about ripping that E-brake (or doing it the janky way with your foot/pedal brake) and revving your engine until your tires spin on the concrete.

Not only are you putting yourself and others in danger, but you are basically not just burning the rubber on your tires, your burning the money in your wallet.

Every time you burn rubber, just picture dollar bills flying from your tires into the hands of happy mechanics and tire salesmen all over the world.

Besides not burning rubber, you should also avoid pot holes and running into or over curbs, as it all adds strain to your tires and suspension parts. One myth that most people fall into is they think the only thing burning rubber (and potholes and curbs) does is wear their tires a little bit. WRONG. Besides getting new tires, this also adds strain to the other suspension parts of your car, which can result to $100s and $1,000s of unnecessary repairs and purchases of new parts in the future.

So when you see someone burning rubber, don't get mad at them. Instead, laugh it up, because you very well know that they will be paying much more in the future for car repairs. And then you can sit back and relax, as you know that by following this simple tip, you are saving yourself money.

9

Ever Get Stuck?

It happens, especially in mud and snow. However, getting your car stuck can happen anywhere.

When most people get stuck, they get angry - I know I do. However, when you get angry, it only makes the problem worse.

Most people try to ROCK IT, meaning hitting their car into reverse, then into drive, then reverse, then drive, and continue to do this until they can get out of their rut.

Lots of things can happen, and one of the worst is excessive heat to your engine. Even though you aren't moving anywhere, your engine begins to heat up fast, which can damage your transmission, wheels, clutch, differentials, and all sorts of other parts of your car.

Also, when people rock their car back and forth (drive to reverse shifting), the wheels of the car spin excessively, which can cause unnecessary damage to your tires. Also, let your wheels spin long enough, and you'll just dig yourself into a deeper hole, which is never good.

If you really want to do this yourself, first clear a path in front of you. Get rid of all the extra snow or mud or crud in front of your tires. Then, add some traction. Gravel, sand, even cat litter! These all help in giving your tires traction to get out of the hole you're in. Once you've done this, it's important to not spin your tires, and start with a slow, steady movement forward. If you have a couple of friends with you, it makes it easier if they push the car from behind. You might not think this helps, but trust me, any extra help without causing damage to your car is worth it.

Finally, if you are REALLY stuck, call a tow truck. Sure, it may cost you some money up front, but it'll save you 100s and 1,000s of dollars in repairs (both immediate repairs and future repairs).

Jim Gaines

10

Keychain Anchors

This is something I was known to do for years when I bought my first car. I call it... The Keychain Anchor.

You've probably seen it before. Either you or one of your friends or family members is "one of those people" who think their keychain should have a bunch of bling and extra crap on it. BAD CHOICE.

Want to know what I used to have on my keychain? My car key, my 3 family members' car keys, 4 house keys, my Public Storage keys (yes, both of them), a Master Lock AND it's 2 keys, a USB thumb drive, a Disneyland keychain, 3 other keychains, tons of circular keychain loops to link all of them together, and a long keychain holder/band thing. Now I laugh at what I used to have on my keychain.

Here's the problem: If your keychain anchors your key down while in the ignition, it puts lots of unnecessary strain on the ignition. This leads to your ignition tumblers wearing away, and can cause your car to one day not start at all. As soon as I found out my ignition was getting loose, I removed everything except my car key.

It's okay to have 2 separate keychains. Just have one for your car only, and one for all the other crap. Or, if you really need the extra stuff, at least get rid of the duplicates of keys, and definitely get rid of the keychains. Sure, it can add some bling to your keychain holder, but it's really not worth the $400+ repair bill.

11

Finding The Best Car Insurer

Finding the best car insurer is key!

And when I say "best", I don't mean "cheapest". So many people fall into the trap of going for what's called 'cut rate insurance'. There are also many other traps you can fall into, here's how to avoid them.

First, don't blindly insure your car with a company just because you saw a funny commercial of theirs on TV. Do your research. When looking for the best, you need to make sure they are reputable. They need to have a good reputation for complete claim payout and fairness.

The worst thing you can do is insure your car with someone, and then when someone crashes into you, they say "Oh well... you aren't a safe driver anymore, so you don't qualify for our extra goodies we promised when you signed up with us." I've had dozens of friends complain about car insurance, and what do I do? I laugh. Well, that is, after they've cooled down a bit. Then I ask them: Did you do your research? And then they laugh at me, and say stuff like "this isn't a term paper, dude."

Ask your friends and family about their car insurance troubles, everyone will have a story to tell. Then, search online on car forums and ask/answer sites to find out what other problems people have had with car insurance agencies. You'll definitely come to a clear conclusion and answer to your insurance problems.

One more thing: Don't pass up on a car insurer just because they are "too expensive". Of course insurance costs money, and the good ones cost even more! Now, I'm not saying the most expensive is the best necessarily, but usually that IS the case. Put it this way, would you rather spend a little extra money per month to be sure you get full claims and forgiveness and CASH BACK, or would you rather spend as little money as you can, and then find out that you have to give up your totalled car and pay for a new one while only getting the minimum from your insurer, having to pay for most of it yourself? You be the judge.

Okay, I left one thing out, and it's really important: ASK THE HARD QUESTIONS. Seriously, write down a complete list of those difficult questions for your 'insurance

Jim Gaines

interviews'. Turn the tables on them, and never make them feel like they are asking all the questions while you blindly answer, as if you are in a job interview. Ask them things like: "What if someone crashes into me? What happens then? What if my car is stolen? What if my car stops working? Will my insurance cover this/that? What if it wasn't my fault? How do you determine who's fault it is?" Ask them the hard questions, and study how each insurer answers them. In closing, you need to make sure they are friendly, have a good reputation, and also make sure you aren't being bamboozled into insuring your car before you have all the facts straight.

67 Need To Know Tips To Extend Your Car's Life

12

When Storing Car, Gas Up

If you aren't going to be using your car for long periods of time, and are storing it away (either in your garage, in a car storage facility, in a public garage, or even in the driveway or on the street), make absolultey sure you fill your tank with gas to the top.

You may be thinking, "If I'm not going to use my car for a week/month/year, why should I spend $50 or $100 or $150+ on filling up my gas tank?"

Well, first of all, filling up your gas tank isn't just about driving it around and burning the gas away. This also adds other benefits, especially when storing your car for longer periods of time than usual.

If you have close to an empty tank, condensation will begin to accumulate in the gas tank. This is horrible for your car now, and even worse for your car later when you decide it's time to take it for a drive. The fuel in your gas tank (the little amount you have left) will begin to evaporate, which causes sludge to form in and around your gas tank. Then, when you decide to take your car for a drive, you're basically running waste, crap, poo through your engine.

As you can already guess, this puts more strain on all the parts of your car, especially the imporant ones under your hood.

Besides gassing up your car, you can also add a bottle of gasoline conditioner to your gas, which will help evoporation to NOT happen. Quick tip: When using gas conditioner, make sure after you add the conditioner to run your car for 5 to 10 minutes to allow the conditioner and the fuel to mix and distribute throughout your fuel system.

One more tip for gassing up while storing your car: Even if you aren't going to be driving your car for awhile, you should still USE it. What I mean is, at least once a week (if able), start up your car, and lightly tap the gas pedal to not only allow fuel to distribute through your car, but also to keep your engine and car parts warm and viable. I've heard the story too many times from mechanics where someone tows their car to his shop saying their car isn't working or makes funny sounds, and the cause of the problem was that it sat for way too long, causing some parts of the car

Jim Gaines

not to work.

13

Before Storing, Wash AND Wax

Washing and waxing your car isn't just to make it look good for yourself and others, it also can do wonders when no one sees it at all (and will hurt your car if not washed and waxed properly).

When you let your car sit in storage, you must wash your car from top to bottom. Don't leave out anything, and don't just rinse your car. If you let it sit without washing, the exterior will begin to attract dust, dirt, and other common crud and mildew that when built up over time, could take hours to clean (or worse, the need for professional washing/waxing/paint refinishing).

Wash the wheels, wash the paint, wash the windows, the metallic areas too. Everything needs to be washed thoroughly in order to preserve your car correctly. Once you are done washing, give it a complete dry - use CLEAN towels to make sure you don't add to the dirt and dust after you just spent the time to wash the entire car.

Once your car is washed and dried, it's time for waxing. Yes, I know, you might think wax is only for show cars and only used by the rich. Wrong. If you really want to preserve your car and make sure it is completely spotless when you retrieve it later, don't skip out on this. Don't skip any area of the car - every area that is painted should be waxed, period.

Also, if you have any chrome surfaces on your car, be sure to polish them with a good chrome polish in order to prevent your chrome from rusting.

Jim Gaines

14

If Your Car Sits For Awhile

You're probably thinking that there are a lot of tasks to do to maintain your car when it sits for awhile. Don't worry, there's just a few more tips to cover, and then we'll get on with other strategies and tips for extending your car's life (when it's not in storage).

Anyway, one thing that a lot of people don't think about when storing their car or letting it sit for awhile is the parking brake.

It's just common habit to rip that parking brake up every time you park your car, but when it's going to sit for awhile, DO NOT engage the parking brake.

Now of course, if you are parking your car on a hill, or somewhere that isn't a flat surface, you can leave the parking brake engaged, in fact you should leave it engaged to prevent your car from rolling down hill.

But, when your car will sit for awhile on a flat surface, keep the P-brake disengaged. If you leave the parking brake engaged, the brake pads may become fused to the rear brakes, especially in cold weather. Also, if the parking brake is engaged for too long, it can begin to corrode, and that's the last thing you want when you start your car and begin to pull away.

If your car is not on a completely flat surface, it is recommended that you place blocks in front of and behind the wheels of your car so that it doesn't roll. Better to be safe than sorry. Also, the next tip in this book explains an easy way to fix this even further.

15

Jack Stands Work Wonders

Besides disengaging your parking brake and putting blocks in front of and behind the wheels of your car, you can also take this a step further.

This tip is what we call jack stands. Of course, you want to make sure you are on a level surface - seriously, even a small incline isn't good for jack stands, so make sure you are on a flat surface. Besides a flat surface, make sure the ground beneath the car is firmly supported (cemented, paved, avoid loose gravel and dirt).

Raising your car onto jack stands is easier than you think. In fact, it is usually stated in your car owner's manual.

First, find the curb weight for your vehicle, and get jack stands that are rated for (can support) the curb weight of your car. Then, engage the parking brake (you will disengage later) so that your vehicle doesn't roll while it is supported.

Next, jack up the vehicle using either the standard jack and wrench that came with your vehicle, OR you can use what is called a "floor jack" that can lift all vehicles effectively. Again, take a look at your owner's manual, most of the time these jacks can be found in a secret compartment in your trunk or underneath your car.

TIP: DO NOT put your jack stand under the floorboards of your car. I did this once, and the jack broke right through the floor of my car. Not a pretty sight. Instead, use the jack on the frame of the car, which is the most sturdy area.

Use your jacks to raise your car on 4 areas - both 2 sides of the car, and on those two sides, front and back (basically just the inside area of the wheels). Each time you raise a section of the car, put a jack stand next to the jack, and lower your car onto the stand. Do this for all 4 areas, and your car should be completely lifted off of the ground. Once your car is lifted, you can then safely disengage the parking brake.

This will not only help your car by having the parking brake disengaged, but will also take much needed weight off of the suspension of your car, including the wheels and tires. Your car will definitely last longer, and you won't have any problems when you go to start your car when you take it out of storage.

Jim Gaines

16

Draining Your Battery

"Drain my battery, why?"

Believe it or not, even when your car battery is not in use, it still discharges over time. And if it discharges too much, it will age before it should. Other problems include battery corrosion, as well as complete battery failure, causing you to have to buy a new battery each time you store your car.

You don't want that, and the way to fix it is by draining your battery. However, I can't just tell you to 'drain your battery' without telling you the proper way of doing it.

There are a few ways to go about draining your battery.

One way is to first disconnect and remove your battery, then every week or so to drain the battery with a small light bulb, then use a low volt charger to recharge it. However, if you aren't the brightest or most comfortable with cars and electric work, you may want to use one of the next draining strategies instead, as it takes less work.

The next way (two different kinds) is the trickle charging technique. Do a search, or visit your local auto parts store, and they should help you find what you are looking for. What a trickle charger essentially does is trickle the charge of your battery, charging it a little, then turning off, then charging it a little, and so on (pretty self explanatory). If you leave your battery on a basic charger, and it sits at the highest charge for long periods of time, it can ruin the battery.

Another term and type of trickle charger is what is called a 'battery tender'. This floats a battery charge at a voltage (not the highest) so that it doesn't sit at full charge, and never lets the battery die out.

Go ahead and invest a little money in a trickle charger, and you'll save the money you would have spent on countless new batteries for your stored cars.

17

Plug Your Tailpipe

Again, this is one of those tips that you may scratch your head at, but trust me, it's necessary if you don't want future problems arising for your vehicle.

Before I go into this, DO NOT plug your tailpipe while using your car. That's a recipe for engine and fuel system disaster. Again, this should only be used when storing your car while not in use.

Plugging your tailpipe helps your car in a couple different ways. First, it will prevent critters and bugs from getting into your fuel system. This is bad, but not as bad as the next possible outcome of leaving your tailpipe unplugged, which is water. And no, I'm not talking about rain water or your friend's playing a trick on you, I'm talking about moist air, condensation if you will. This especially happens in the winter, or when it is rainy and cold out (but can really happen at any time). The last thing you want is water in your fuel.

So what should you use to plug your tailpipe?

Well, really anything will do, I've used dry rags or towels to plug my tailpipe. Many auto technicians recommend steel wool. If you really want to avoid hassles of bugs, critters, and moist air, you should plug any other orifices of your car that these may affect.

TIP: When you are ready to take your car out of storage, make sure you remove all pipe-plugs. Sometimes your car won't start at all, and sometimes it will start and then blow on you minutes later. Of course, after long storage periods, you should always check EVERYTHING before taking your car out.

18

Parking In The Shade

Okay, this is the last tip for storing your car (of course other tips in this book can be used for storage techniques as well).

This should go without saying, but you'd be surprised at how many people don't know (or don't care) about this - but it's essential to your car's health, and we all love healthy cars.

Park your car in the shade.

That's right. Whether it's in a garage, in a storage unit, car cover, windshield cover, it all helps!

I don't know where you are going to store your car. However, people store their cars in all sorts of places. If you store your car in a storage unit where it's JUST your car inside a clean, air-tight unit, you should be fine.

But then, if you store your car in a garage with all sorts of other crap in it, I suggest putting a car cover over your car. This also goes for parking your car outside. And if you don't want to invest in an entire car cover, at least get a windshield cover.

Parking in the shade, indoors, and covering your car help preserve your car from all elements. The sun, dirt, grime, water, condensation, air, critters, and many other things you don't even think about could affect your car's health.

19

Properly Clean The Interior

Okay, now on to the general car care tips. First, we'll start with your car's interior.

90% of cars I see/work on/been in have worn seats, stuff everywhere, and dirtied floors. Now, if these people simply took care of their interior like they do their exterior (sometimes), problems like corrosion and interior damage wouldn't occur, ever.

Have you ever heard the excuse, "Car's wear over time, it just happens..."? Well, that's exactly what it is, an excuse. An excuse for their lack of taking blame and ownership in their own faults.

First, get rid of the extra crap in your car, it's not a storage unit. The more stuff you have in your car, the more your seats wear, the more your carpeted floors get ruined and scuffed, the more "general wear" happens.

Next, you want to make sure that every time you wash your car, you clean the insides as well. Clean, vacuum, wash, and dry the insides to avoid corrosion and damage.

Use a combination of detergent and water to wipe dust from every inch of your interior. Pull up the seats, get all the little cracks that tend to wear away first. Get the lenses, the dash gauges, the insides of your windows. The more thorough you are, the less of a hassle it is to clean later, and the more prestine your interior will look.

Trust me when I say this, not everyone's interior wears. In fact, I've seen cars that are 10+ years old that have flawless looking interiors. How? They took care of their interior. Not just once, but all the time. Take pride in your car, and take care of it.

Jim Gaines

20

Get The Very Best Floor Mats

This goes along with the previous tip about taking care of your interior.

Most cars come with floor mats, some don't. If your car doesn't have floor mats, GET SOME. Even if your car has floor mats, unless you bought a really expensive car, there are better floor mats out there.

Floor mats not only help the floors of your car with slush, mud, salt, and dirt, but also gets rid of the wear and tear that can happen. How many times do you get into your car every day? How long do you spend in your car? Unless you are driving around with no shoes and socks, the floor of your car gets beat to hell. Scuffs, tears, rubs - It doesn't seem like much now, but give it time, and your floors will be a mess.

One type of floor mats that I buy for all of my cars are the waffle-style mats. These literally stick to my floors, and don't slip around like other common floor mats do. Also, they are extremely easy to vacuum and wash off.

By getting some good waffle mats, you will protect the interior floors of your car, and will save time and money when you clean the mats. This should really be a no-brainer.

21

Preserving Door and Window Seals

This is another one of those "things" that people don't think about, and could really care less about. However, it's also one of those "things" that could cause lots of unwanted problems for your car in the future.

When you don't preserve your seals (usually for doors and windows), it can cause a lot of problems, the most common being leaks. Yep, in my old Honda, I had a problem with my window seal, and one rainy day, the entire car filled up with rain water, and I had a hefty bill to replace most of the inside of my car. Unfortunately, I didn't have the money to spend, so I was forced to live with the disgusting smell for months.

Window and door seals can dry out and become cracked, and eventually will dislodge themselves from what they are protecting if not preserved.

First, every time you clean your car, both interior and exterior, make sure you use soap and water to clean your seals. Get rid of the dust and dirt that builds up in and around these seals.

Next, you can use armorall, or any silicone based product on the window and door weather stripping to keep them conditioned. This helps them stop from drying out and cracking away. I remember one time I asked my buddy where his weather stripping was, and he didn't even know he had weather stripping! Apparently, he never conditioned the seals, and the stripping dried out, cracked, and fell off of his car.

TIP: NEVER use oil-based products (like WD-40) on seals. Oil based products can damage your rubber, and will do just as much harm as not doing anything at all. Soap and water works best, and to take things further, use a silicone-based product on your seals to ensure their stability.

Jim Gaines

22

Cracking Leather

This is common for cars with leather interior (even the fake leather, or "fleather" as I like to call it). Sometimes cars don't have leather seats, but other parts of the car (like door panels) are leather.

And again, people come up with that same excuse, "leather cracks over time". But no one finishes the statement. It should be, "Leather cracks over time, when not taken care of by it's owner."

If you have leather in your car, use leather cleaner to stop it from drying and cracking. Leather cleaner will remove dirt and dust from your seats, and will keep your interior looking clean and flawless.

Besides leather cleaner, you can also follow that with a leather protectant to resist stains. Next time you or your friend spills something on your seats, it will be a much easier clean up process, and you won't have stained or grimy leather afterwards.

Taking care of your leather now also makes it easier for future cleaning. You don't need to use leather protectant every time you wash your car. Just use it every once in awhile, and you will find that it's easier to clean your interior with just soap/water and rags in the future. This will also keep your leather from fading, cracking, drying, and other effects of time.

23

For Carpeted Interiors

Of course, most people don't have the luxury of having a leather interior in their car. Most of us have carpeted interiors.

Although carpeted interiors don't really 'crack' like leather, they have all sorts of other problematic things that can happen. Like, it's much easier to wipe away grape soda from a black leather interior than it is to clean from a carpeted interior.

So, how in the world can you truly protect your carpeted seats?

Well, to start, you can apply any home upholstery cleaner to your seats. Just rub it on in there, and wipe it off with a clean cloth. Right now, Scotchgard is a popular cleaning agent/product that I've used before, but any upholstery cleaner works. If you have the money to invest in car-specific carpet cleaners, that works even better.

Not only do these clean your carpeting, but it also helps to resist dirt and dust from building up, as well as helps to reduce stains that can happen in the future.

Also, besides using cleaning products on your interior, it's recommended that you have seat covers for your seats. No, not the $100+ expensive ones. Just simple cloth-style seat covers that cover the part of the seat that your head and back rest on. These are usually inexpensive, but can work wonders for your car. They reduce stains, wear and tear, and next time you spill something on your seat cover, you can take it off and either replace it, or smile at your flawless looking seats.

Jim Gaines

24

Cover Car From All Elements

I touched on this topic in an earlier tip/technique, but that was more for car storage. This goes for everyday use of your car.

This goes with protecting your exterior paint, your interior paint, help from fading, interior leather and/or carpeting, your headlights/taillights, and other common areas of the car that you want to protect.

When the sun is blisteringly hot, it can cause your paint to fade. And when magnified through your windshield and windows, it can cause your interior to crack and fade as well.

The first step to protecting your car from the sun and other elements is to try to park in the shade as much as you can. You don't need to park directly under a tree, but just try to park in a place that isn't directly under the sun.

When your car is at home/in front of your house, use a car cover to cover it from the sun, the rain, dirt and debris.

When you are traveling, or are at work, or somewhere where covering your entire car isn't viable, use a simple windshield cover to protect your car's interior, as well as keeping the inside of your car from getting too hot.

Another thing that I do to keep my car from getting hot (as well as the sun's rays from damaging my car interior) is tinting the back windows of my car. This goes for the rear window, as well as the back two passenger windows. In my state, it's illegal to add any tint to the front windows, so check your state laws to see which film tint is legal.

The point is, it's very easy to protect and cover your car from the elements, and to avoid lots of damages (minor and major) from occuring in the future.

25

Wash Car Not To Look Nice

We talked about washing your car before storing, but it's also obviously important to wash your car when in use.

And not only should you wash your car to make it look good, but you should also wash your car to prevent damage from your paint finish. "Oh, is THAT why my rich neighbor washes his car every day?" Haha, I'm not stating reasons for other people, but it's important to wash your car, let me explain.

When you drive your car (as opposed to letting it sit), even though you aren't as likely to have problems with dirt and dust building up over time, you ARE more likely to incur other damages on your car. Sand, road salt, slush, dirt, dust, water, acid, and other common things like this happen every day when using your car. Some of these things (like acid and road salt) you don't even notice and can't see.

When you wash your car every week or two, it helps to get rid of ALL of this crud. If you don't wash your car, you are allowing all of these elements to build up, which will more than likely make your paint fade, chip, and become damaged in some way.

One thing you can do a few times a year (I do it every 3 months) is use buckets of warm water to wash your car with (make sure the temperature is above 0, or not freezing, as this could crack your windows). Either pour the entire bucket on your car, and use a sponge and car-detailing brushes to wipe the warm water around, or just dunk sponges into the warm water, and wipe your entire car down. This helps to remove slush and road salt from your car easier. When you do this, you should also use tire and rim cleaner to get rid of slush and road salt from your wheels as well.

Jim Gaines

26

The Very Best Type of Wax

Okay, I've already previously told you about the important of waxing your car, especially when storing it for long periods of time. Well, besides storing your car, waxing is a good choice every time you wash your car.

Not only does waxing give your car a clean, shiny look, but it also protects your car from acid rain, road salt, slush, and the general deterioration of your exterior paint. I recommend waxing your car 100%.

However, all wax is NOT created equal. Most people who decide to first wax their car will go to the nearest auto parts store, and go to the "wax" section. Then, they browse up, down, and all around, and decide on their wax based on the price. DON'T DO THIS.

There are two different types of wax that most people use on their cars. The first is liquid wax. This is the type of wax that pours out of the tube onto your car. Although this is sometimes easier than using other types of wax, it is also the worst type of wax. It doesn't protect your car like this next type of wax I'm about to tell you about does.

The type of wax that I as well as professionals use is paste wax. Paste wax usually comes in a can, and instead of pouring it out, you use the supplied applicator pad to rub the wax onto the pad, then proceed to rub around your car's exterior paint. Paste wax is stronger, harder, and lasts so much longer than liquid wax.

Don't apply this wax too thick, or it will become very difficult to remove. Just make sure you apply it evenly across your exterior paint. Once you have waxed an area of your car, use a soft cloth to remove any dry wax residue that you may have left behind (to avoid scratching your paint).

WAXING QUICK TIP: I always apply 2 coats of wax to the hood of my car. Why? Put simply, the wax on your hood will wear away much quicker due to the heat of your engine. If you drive your car more often than others, your wax on your hood will wear away even quicker, so apply the extra wax to your hood and protect your car's exterior!

27

Self-adhesive Car Skins

This may not be your cup of tea, but nonetheless, this SAVES the exterior of cars completely.

What I'm talking about is self-adhesive car skins. Washing helps protect your car from dirt, dust, salt, slush, and general crud in the air - so does waxing. However, there's not much that just washing and waxing will do for stones, pebbles, and other larger crap that could fly up and chip your paint. Your car will still look clean and shiny, but the chip in the paint remains no matter what.

You can protect your car from most paint chips using a self-adhesive urethane film.

If you are a handy kind of person, you can try applying the transparent film to your paint using Scotchgard urethane films, I think 3M also has one. But of course, look around to find the very best. Once you apply the film to your car, you can wash and wax and drive around like normal.

However, this is one of those things that may take some professionalism in order to get right, so I recommend you find a body shop that can do it for you. And I'll say this a million times in this book and list: Don't go for "the first", and don't go for "the cheapest". What I mean is, really do your research when trying to find a reputable business to care for your car. Make the mistake of going for the first you find means you may pay a lot more upfront, and make the mistake of going for the cheapest, and you may find yourself paying for unnecessary repairs and "re-dos" later.

Also, while you might feel this is an unnecessary addition to your car, think again. And don't just think, "I don't need my car to look great, it's okay if I get a chip or nick or two in the paint". It's not just about the looks of your car right now - think about when you sell it. If you have faded paint, nicks and chips in the paint, and other rubs and 'bad areas' of the car, it lowers the value tremendously when going to re-sell it. However, if you have a flawless looking car, it's 10 times easier to sell, TRUST ME.

Jim Gaines

28

Touching Up Paint Nicks

In the previous tip, we talked about how to PREVENT nicks and chips from happening. But of course, not everyone is perfect, and accidents happen, and I'm sure you or someone you know has at least 1 or 2 chips in their car's exterior.

This tip is for all of you, the people who couldn't prevent it, and now need to fix it.

If you get a chip in your paint, it's time to touch it up. You know what happens to little paint nicks and chips? The area that is chipped becomes rusted quickly. Then the area around it begins to wear away, and the rust spreads. And spreads.... And spreads. Before you know it, you have an entirely rusted area of your car. I've seen some old cars get completely rusted in just a year or two. When this happens, you can kiss any chance you thought you had to sell your car goodbye. The only thing you'll get is $100 or $200 from the junk yard to get rid of your rusted piece of metal.

It's very easy to touch up your paint chips and nicks. Obviously, you can pay someone to do this (it's not too expensive). But, I see no sense in that because of how easy it really is to do it on your own.

First, you must get rid of any rust. You can do this with rust remover, or rust arrestor, to stop rust from forming underneath your paint (which can spread around your entire car).

Next, sand the area of your car that has the paint nick. Use a small piece of sandpaper, and make the area rough so the new paint adheres correctly to your car. Then, wash thouroughly. Wash away the sanded paint, rust, and other dirt and debris around the area you have just sanded.

Now, if you don't want to apply new paint, you don't have to. All that you have just done will prevent rust from moving to other parts of your car. However, if you leave this exposed, you will have to go back and remove the rust over and over again when it comes back. So, I'd suggest continuing with the primer.

Use a SMALL brush (sometimes I use a toothbrush, or a very small artist brush) to get the primer in the previously rusted/nicked area. DO NOT get any primer on your car

67 Need To Know Tips To Extend Your Car's Life

paint! If you make a mistake, wipe it and wash it off right away, or you may have a partially raised/blotted area of paint on your car.

Nowadays, you can find an EXACT paint color match to your car. Do 5 minutes of research and you can find the exact paint needed for your exact model/make/year/color of your car. Apply the paint to the area, starting with the outer edges, then moving in to the center of the nick. Make sure you apply the paint evenly, and again, don't make it too thick, or it could look blotchy and raised.

Once paint has been applied smoothly and evenly, wait at least 24 hours (I recommend 48 hours). Then, wash, wax and polish the entire vehicle to blend the new paint with the old paint. This brings your car to a high glossy state, and makes the entire car look like it was just completely painted.

Jim Gaines

29

Easy Light Cover Repair

First of all, let's talk about foggy headlights.

If you notice your headlights are getting foggy and oxidized, it's very easy to fix. I've actually found that those 'Dore Does It' rags and other similar rags can do the trick. But sometimes, the oxidation is too powerful, and the rag won't help at all.

Before going ahead with this fix, find out if the oxidation is occuring on the inside or outside of the headlight lens. If it's on the inside, you will need to remove the headlight lens/cover, and continue the following steps on the inside instead of the outside.

First, clean your lens with a solution like Windex or something similar. Any type of degreaser works, but I'd recommend watering down the degreaser before wiping your lens. Then, use car polish (or plastic polish) right on the lens, and wipe accordingly. Make sure you don't get ANY of these solutions on the rubber or painted parts of your car. If you do, wipe it off immediately. Finally, use some sort of buffer (or just a soft rag) to wipe and clean the crud from your headlights.

Now, some people use sandpaper to get rid of oxidation. Although this works faster, it is also prone to mistakes. Scratch the surface of your headlights too hard, and you could scratch your covers, forcing you to buy new ones. However, if you DO go the sandpaper route, make sure to dip the sandpaper in a soapy water substance before wiping. Then, use a watered-down degreaser and/or polish after sanding like before.

Now let's talk about broken taillight covers, you know, the red and orange taillights.

Before just bringing it into the shop and saying "fix it", find out the different solutions to your problem. You may not need to replace the entire taillight, it may just be as easy as the cover, which shouldn't cost much at all.

If the cost to repair the taillight is too expensive for you, an easy fix is to tape up your taillight. This not only helps to prevent moisture from occuring inside your taillight, but also makes sure you won't get pulled over by the authorities for having a broken taillight.

67 Need To Know Tips To Extend Your Car's Life

There is special orange and red tape that is made specifically for this problem. It also adheres much stronger than any old normal tape. You can decide to either use the red/orange taillight tape only, OR you could do it the way I do it. First, I tape the outside edges of the taillight cover with basic masking tape. This helps to prevent rain and water from building up in the sides that could have shifted around when the cover was broken. Then, I use clear packing tape to cover the area that was broken. This gives an extra bond that prevents water as well. Once I have done that, I cover the entire light with red tape (red or orange depending on the part of the light you are repairing).

This gives you an easy, inexpensive fix to your broken taillight cover problems. Of course, if it gets even worse, and water build up inside the taillight, it could cause your taillight/brake light to not work, which is a common way to get a "fix-it ticket". So, keep a look out each time you get in your car to make sure your taillight is still working properly.

Jim Gaines

30

Properly Change Your Bulbs

When I was a kid, about 17 years old, I didn't know much about car repair. So, every time one of my headlights or taillights went out, I'd drive down to the auto shop and pay them a good $20 - $50 to "fix it". Then, when I got more into car repair, I found out all they were doing was simply switching a single bulb that cost $5 or less, and taking just 5 minutes to do the complete fix.

After I realized how much money I was wasting, and how much time I was wasting waiting for these mechanics and auto-repair shops to 'get to my car', I decided to figure out how to do it on my own. And man is it easy.

Did you know that with most taillights, you don't even need to remove the taillight to fix the bulb? There's usually some switch or compartment on the inside of your trunk that you pop open, and WHA-LAH! There are all the bulbs for your taillight.

First, find out what bulb is not working. Then, twist (or pull, depending on the car) the bulb until it is removed, and look at the tiny letters and numbers on the bulb. This will tell you what bulb you need to replace it with. Next, go to an auto parts store, and smile when you find out you only have to pay a few dollars to get a replacement bulb.

Then, simply go back inside your taillight/headlight (however you removed the bulb), and put in your new bulb.

Before you go driving away, MAKE SURE all your lights work properly. And I'm not just talking about the light you replaced.

TRUE STORY: I once replaced a bulb, and because it was the wrong type of bulb, it caused ALL of my lights to malfunction. When I put my car in reverse, the brake lights came on. When I would brake, the reverse lights came on. My indicator lights stopped working entirely. It was a mess. And it was all because I had the wrong bulb (in fact, it was the correct bulb really, but instead of a "3", there was a "6", which caused more power to be sucked from the other bulbs, causing malfunctions in my entire lighting system).

67 Need To Know Tips To Extend Your Car's Life

Now that you know how easy it is, GO DO IT! Then, when your friends and family need help with their bulbs, tell them you'll charge them HALF what others do, and that you'll do it in HALF the time. Or, you can just laugh at them and tell them you replace your own bulbs for a tenth of the price.

Jim Gaines

31

Small Windshield Chip Restoration

Everyone has been through it. You're driving along the highway when all of a sudden, SMACK! Your windshield was just hit by a rock, and then you find out it was because of some trucker in front of you (you probably have a similar story).

It is very important that you restore the chips in your windshield before they get worse. Besides the chip looking bad, it also causes distractions when driving. Also, water and dirt can get into the chip, which messes with the windshield glass, making the next pebble that hits your car turn your chip into a large crack. The last thing you want is to have to replace your entire windshield.

I actually use Safelite (yep, their commercials got me hooked, 'Safelite repair, Safelight replace!'). They actually have a deal with my insurance company, so every time I get a nick or chip in my windshield, they use their "special resin" to repair the nick without having to replace the entire windshield - for free.

However, if you don't have a service like this around you, there are actually little 'mini-kits' for windshield repair. I've seen them all over the place, and they have never been too expensive. Get a hold of one of these repair kits, and follow the instructions to fix your windshield nick/chip.

In a nutshell, the windshield repair kits work like this: You use their tool to suck the dirt, moisture, and little broken glass particles out of the crack. Then, you clean the area of the chip. Finally, you "inject" a glass resin that basically combines into your windshield, making it as if the chip was never there. Trust me, repairing the little chips and nicks in your windshield is worth it. Wait around too long, and you'll have to usually pay hundreds for an entirely new windshield.

32

Before Hauling Stuff on The Roof

Like I stated before, some of these tips are very short, sweet, and to the point. This is one of those tips.

Before you haul a bunch of crap on the roof of your car, it's recommended you take the necessary precautions.

First, check your car owner's manual for vehicle specifications. Smaller cars won't be able to haul much on their roof. Could be 100 pounds only, most are around 200 pounds, while only some cars can support more than that.

If you haul too much on the top of your car, it can cause the weight distribution to change dramatically. Your car was built with specifications. Change the weight distribution, and the specifications go out the window. Take a turn to quick, and your car could flip. Go over a bump or pot hole, and the top of your car could dent downward (VERY expensive repair).

I also recommend that whenever you haul stuff on your roof, first place a piece of carboard or a blanket along the entire top of your car. This will prevent your car from scratching, or worse, denting.

Jim Gaines

33

Securing Your Big Loads

Going off of hauling stuff on your roof, this is for big loads that you add to your car - they can be on the back of your car, inside your car, on top, on the sides, ANYWHERE.

Like you learned before, it's always best to secure a blanket or piece of carboard wherever it is you are hauling stuff to prevent cracks, dents, scratches, and chips.

But besides that, it's always necessary to make sure you secure your load. I've seen it a thousand times, people not securing their loads on their cars, and while driving, their entire load falls off. This is an inconvenience for them, and a death wish for others around them. Don't let this happen to you.

I ALWAYS recommend cargo straps. Yep, the ones shipping businesses use, the ones they use on planes. Heavy duty stuff. Get a bunch of these, and make it easy on yourself while securing your big loads.

Although, if you don't have cargo straps, at least use rope (NOT twine), or even a good strong bungy cord of some sort.

If you have the money, invest in a luggage rack, bicycle rack, cargo rack, or some sort of compartment you can attach to your car to put your big loads into. On one of my old SUVs, I got a "Rocket Box" (I think it was from Yakima at the time, it was years ago though), and any time I had extra crap to haul, I'd throw it in there, lock it up, and BAM! Secured!

34

Wheel Well Splashguard Inspection

Did you know these things even existed? They are called wheel well splashguards, and they are installed/come with every single car I've seen or worked on.

Wheel well splashguards are designed to keep water, slush, and other crud from splashing up into your engine and other parts of your car. If your splashguards are damaged, hanging by a thread, or have fallen off, water can get into the electrical components of your engine, which can cause MAJOR problems for your car.

So, every one in awhile, check and inspect your wheel well splash guards. Make sure they haven't fallen off. But even if they are there, make sure they aren't damaged, cracked, or loose. Most of the time, these splashguards are flimsy, and can rip and tear off easily, and you'd never know unless you checked.

If these are loose, you can refasten them. Depending on the make and model of your car, you can use the clips that come with it, or sometimes you can just superglue the area that tears, or use some sort of zip tie to fasten it to your car.

If you can't seem to refasten it, you will need to replace the splashguard. Search your make and model of your car to find out what it will cost you.

Easy Replacement Tip: I ALWAYS visit my local junkyard or car impound lot for car parts. Splash guards from the junkyard won't cost you much at all, and you can find the exact ones you need for your car. Give it a try, and remember: ALWAYS inspect and check to make sure your splash guards are properly fastened and doing their job.

… Jim Gaines

35

How To Check Uneven Tire Wear

Sometimes, you can just FEEL that your tires are uneven while you drive. But sometimes, if the problem hasn't gotten that bad, you will only be able to notice uneven tire wear when you look at and inspect your tires up close.

Uneven tire wear can be caused by TONS of different things, including inflation problems, rotation, alignment, worn shocks or components, balancing, and many others.

Sometimes tire wear occurs on the edges of your tires, sometimes it's in the center, and sometimes it can happen at random areas of your car. Uneven tire wear is bad for your car, and doesn't just affect the tires, but your entire car. If your car wears only in one area, it could cause your engine and suspension to work harder sometimes.

Not to mention, it is very unsafe to have worn tires. Let's say you slam on the brakes one day. If your tires are worn in one area (or more), it could cause your car to skid further than you planned, and could cause your brakes to not work effectively when needed.

The next few tips will help you with tires and suspension, but if at any point you notice your tires wearing unusually, or even more than usual, go get it checked out by your TRUSTED mechanic.

36

Tire Tread and What It Means To You

It's different in all countries and areas, but most of this pertains to everyone.

In the United States, there are actually what are called wear bars that are molded directly into the tires, which makes it very easy to notice when you MUST legally change your tires.

However, obviously not all tires are created equally, and you may not have wear bars. Good thing theirs lots of nice little tools and tricks that can help you understand tire tread.

The general rule is when your tire tread has been worn to 1 1/16" (or 1 1/2mm), you should think about replacing your tires to avoid unnecessary accidents and repairs.

Do you know about the Penny Trick? Here's how it works: With Lincoln facing you upside down, place the penny into the tread of your tire at the thickest part. If you can't see the top of Lincoln's head, your tread is fine. If the tread touches the top of Lincoln's head, you will need to replace your tires soon. If you can see the top of Lincoln's head completely, this tells you that your tread is worn WAY TOO FAR DOWN. Replace your tires immediately.

Don't have a penny laying around? At every auto parts store, you can find a tread gauge or tread depth indicator of some sort. I've seen these for as little as 10 cents, and no more than a few bucks. Some of these are so small, you can just attach it to your keychain.

Just make sure you check your tire tread every month or so, and if it gets to small, it's time to get some new tires!

Jim Gaines

37

Valve Caps Are A Must

One of the smallest parts of your car is the valve cap that goes on your tire. These are the little caps you twist off when you inflate your tires with air.

While you might think this is an unnecessary piece of the puzzle which makes up your car, it is actually VERY important.

Let me give you a cause and effect example: You take off the valve caps to inflate your tires. You don't put them back on. You drive away. Dirt and dust and moisture builds up in the air valves. Air begins to leak out of your tires. The more you drive your car, the more air leaks out. You don't notice air leaking, and when you hit a bump in the road, your tire pops. If you are lucky, you will be on the side of the road with a hefty fine for fixing your tire. If you aren't as lucky, you could indeed totally wreck your car, and could not only cause damage to your car and yourself, but the cars AND lives of others around you.

Yep, it's happened before. All because of a simple valve cap. This can also happen with cracked or damaged valve caps, so even if you see your valve caps haven't fallen off, wiggle them around a bit to make sure they aren't damaged, as this can lead to the same problem.

TIP: Tire shops are supposed to provide brand new valve caps when they replace your tires. However, most just use your old valve caps. So, make sure you ask the tire shop to make SURE they put new valves/caps on the new tires. One "secret" trick that I've used on tire shops is: Before I give them my old tires (or entire car), I remove the valve caps. Then they HAVE NO CHOICE but to put on brand new valve caps. Yep, pretty sneaky, but it works. Every time.

38

Keeping Tires Properly Inflated

First of all, this goes both ways. Many think uneven tire wear happens only when their tires are underinflated, but it can also happen from tires being overinflated.

When tread is deeper on the edges of the tire, it's a sign of overinflation. When tread is deeper in the middle of the tire, it's a sign of underinflation.

In order to make sure your tires are properly inflated, first check the writing (usually just embossed in the rubber) in your tire and note the proper PSI. My tires are rated at 44PSI.

You can either inflate your tires at most gas stations, or you can get an air compressor and tire inflation gauge/add-on to inflate your tires at home. No matter what you use, make sure you inflate your tires to the EXACT PSI. Seriously, don't be one of those people that think, "Well, even though my tires should be at 45psi, I'll inflate them to 50psi, since they'll deflate a little over time". Guess what? Over-inflate your tires, and you'll be replacing them quicker than ever.

Besides the gauges that come attached to the air compressors that inflate your tires, you can buy just the gauge for a dollar or two (maybe more), and can easily check your PSI whenever you want.

QUICK NOTE/TIP: When driving in extreme conditions, either very hot OR very cold, your tires deflate faster (this is due to the rubber depressing or stretching depending on hot or cold weather). So, if you have been driving around in very hot or very cold weather, make sure to check your tire inflation more often.

Jim Gaines

39

Got Wet Thumb?

Ahh... the wet thumb test/trick. I love telling people these little tricks, and recommend them as well as you to tell others these little tricks.

In the previous tip, you learned that you can inflate your tires either on your own (using an air compressor) OR you can use the gas/service station air pump/compressor to inflate your tires.

This tip goes with inflating your tires at service stations, and is called the Wet Thumb Trick.

Here's how it works: Before you inflate your tires, press your thumb nail against the pin inside the guage/hose (the little pin that you stick into your valve). If your thumb gets wet, that means there is moisture combined in the air, which will push moisture into your tires - THIS IS BAD, VERY BAD.

If your thumb gets wet, NEVER, and I mean NEVER go to that service station ever again (or just try the test again next time you go). When moisture gets into your tire, it leads to tire pressure variations, which can either overinflate or underinflate your tires. And since you already know what over/underinflation can do to your tires (and entire car), you know that moisture in your tires is a BIG no-no.

Moisture in your tires can lead to unsafe driving, quicker and more frequent tire repairs, rim/cap corrosion, and rusted valves. So, next time you inflate your tires at the service station, do a simple 5 second WET THUMB TEST!

40

Rotating Tires

Remember when I talked about uneven tire wear? Well, one of the biggest causes for this is not rotating your tires regularly.

Each and every car (check your owner's manual) gives you the recommended rotation pattern and periods, and each one is different.

If you were to drive your car and never rotate the tires, you'll notice that your tires will wear in one spot only. This is because of the continuous wear each side of your car produces on your tire. If you never rotate your tires, you can find out that your right tire wears on the left side, while your left tires wear on the right side. This can happen depending on the front/rear sets of tires as well.

If you DO rotate your tires, your tires will wear more evenly, and you won't have to spend money on new tires all the time. Also, it makes your car easier to drive, takes more stress off of your suspension, and avoids any unwanted accidents from occuring.

Since I'm a stickler for safe driving (and keeping money in my wallet), I rotate my tires every 4,000 - 5,000 miles. A lot of the time, I'll just rotate them on every other oil change. For most cars, the recommended rotation should occur every 6,000 - 8,000 miles (9,000 - 12,000km).

TIP: If you rotate your tires for the first time after 5,000 miles, make sure you keep this schedule! If you rotate at 4000, then 10000, then 7000, no matter how many times you rotate your tires, this will cause uneven tire wear as well. So get on a schedule, and STICK TO IT.

Jim Gaines

41

Be Nice To Your Battery

It happens to all of us - you leave a light on in your car, and by morning, the entire battery is completely drained. Here's some tips on how to jump-start your car without doing too much damage to the battery.

First, did you know that more and more auto technicians and car experts are saying that jump-starting your car using another car and jumper cables is becoming more and more unsafe?

So, I always recommend you spend the money to get a battery charger. Keep it in your car, and use it to recharge your battery. This is the absolute safest way to charge your battery.

But, if you are like most, and still rely on jumper cables and a fellow good samaritan to jump your car using their own, here's the correct way to do it:

Make sure both cars are OFF. Connect the positive (red) cable to the positive battery terminal (red) on both cars. Then, connect the negative (black) cable to the negative battery terminal (black) on both cars.

Most people try to start their cars right away. WRONG (and unsafe)!! Instead, wait at least 2 minutes (5 is recommended). If you want to speed up the process, it is OKAY to start the car that you know works. But don't start the car with the dead battery right away - you must wait 2 - 5 minutes for that.

Once the already-working car is started, and has been running for a few minutes, try starting the dead-battery car. If it doesn't work, wait 1 minute, and try again 5 times (waiting 1 minute in between each try). If it still doesn't work, you're going to have to get your car towed, because you my friend, need a new battery.

FINAL TIPS: Once your car starts, disconnect the cables in reverse order. That means take the negative cables off FIRST (black), and THEN take the positive cables off (red). Also, if you aren't sure how dead your battery was before you recharged it, make sure you drive at least 10 miles to completely recharge your battery. True story: My car wouldn't start about a mile away from my house. So, I had a friend jump my car

67 Need To Know Tips To Extend Your Car's Life

with hers, and it worked! Then, she drove away back to her house, and I drove the 1 minute (about half a mile) to my house. Next time I got in my car, it wouldn't start again! I took it to the mechanic (I was young and naive), thinking my car was 'dead', and he told me I should have drove around for awhile after recharging my battery. Apparently, my battery wasn't fully charged after the initial jump, so when I got home and turned off my car, my battery didn't have enough juice to start up again.

42

Wheel Cleaning and Cleaners

Wheels/rims - These are the first thing that get beaten to crap when you drive your car. It's very simple why this happens. Brake dust, dirt, road salt, slush, pebbles, acid rain, and other common factors can lead to your rims corroding and becoming dirt-stained for life - unless you know how to both clean AND prevent this from happening.

Soapy water is a good start. But of course you've probably used soapy water on your rims while washing the rest of your car, and though it helps get rid of surface dirt and dust, it doesn't get rid of the stains and grit and grime that build up on your wheels.

Whether you have chrome rims, aluminum rims, or any other type of hard metal, there is a specific formula cleaner for each kind. Instead of getting your rims professionally washed (usually costs a lot more than a regular car wash, since sometimes this requires removing and reattaching the wheels), you can definitely wash and protect your rims on your own.

Go to your auto parts shop, and get the wheel cleaner formula that is right for your wheel/rim finish (chrome/aluminum/etc.). Then, after washing away all the dirt and grime from your rims that you can with basic soapy water, use the rim cleaner on your car until all the grime has been removed.

It is also recommended that you use wheel polish on metal wheels, as this will prevent and protect your wheels from future grime and wear. And as for all you who don't have metal wheels, and instead have plastic/painted wheels, you can use liquid OR paste wax to remove and prevent grime from building up.

Besides making your rims/wheels look good, getting rid of the crud that builds up on them also prevents the grime from continuing to build up around the wheels, rims, brakes, valves, and other areas around the wheels of your car.

43

ALWAYS Lubricate Your Lug Nuts

I'm going to start this tip of with a cause and effect example: Let's say you don't lubricate your lug nuts (on your wheels) ever. Well, eventually, your lug nuts will begin to corrode, and the corrosion will seize your nuts to the studs, meaning you won't be able to get your lug nuts off anymore. This means you can't change your tires unless you have a mechanic break your lug nuts (which can lead to damaging your tires and wheels, which leads to replacing all of your wheels).

Your answer? - Anti-Seize Lubricant. Every time you rotate your tires, it is recommended that you use this anti seize lubricant on all of your lug nuts on all of your wheels. You can get this at your nearest auto parts store. After cleaning the studs, wipe on the lubricant on all lug nuts. Besides preventing your lug nuts from seizing, this lubricant also stops your lug nuts from slowly spinning off of your wheels. This avoids unnecessary repair costs, and makes your car safer to drive.

Now, if your lug nuts have already been corroded and seized, there is ONE thing that you can try before taking your car to the mechanic. Use a degreaser such as WD-40 (or anything similar, any old degreaser) on the lug nuts that have been seized. Get the degreaser all around the lug nuts, and wait about 15 to 20 minutes. Then, using a ratchet, try removing the lug nut. Depending on how far the corrosion spread, you might have to replace one or more lug nuts. If it still doesn't come off, try the degreaser again. If it's still a no-go, take it to the mechanic.

However, since you probably don't have the lug nut seizure problem right now, make absolutely sure you use the anti-seize lubricant to prevent it from happening in the future.

Jim Gaines

44

Ever Lose A Hub Cap?

If you are laughing right now because you have rims instead of hub caps, stop laughing and just go to the next tip. Although none of my cars have hub caps (one of my old one's did), there are millions of cars that still have hub caps, therefore millions of people will find this tip useful.

Hub caps get damaged over time, and will eventually break loose from your wheel, causing them to be lost forever. Usually, even if you are able to locate and return your lost hub cap, it won't fit back on properly, if at all. Hub caps can also break lose from being installed improperly.

One thing you can do to insure your hub caps don't break lose is to use a rubber hammer or mallet to tap the edges of the hub cap in a circular motion. Don't hit your hub caps too hard though, or you will break the clips that hold it on your wheel!

If you have those old metal hub caps, they are installed by clips that bend into place. If you pry these clips outwards (not too far, just a little), you can ensure that your hub caps will remain on your tires.

Now, if you have a newer car, and it has plastic hub caps, they are installed and held in place with a wire retaining ring. There's not much you can do besides occasionally checking to make sure the ring and wheel tables don't become loose. Once they become too loose, the ring will snap out of place, fall off, and there goes your plastic hub cap, rolling away from your car.

One more tip: You can buy silver/gray cable ties and attach your hub cap with a couple of these (only do this if the other tactics don't hold up). Tie the hub caps tight, and cut any remaining cable ties, and you won't see any excess ties hanging around.

45

Wheel Alignment

It is very important that you not only get an alignment when you get a new car (used car, but new to you), but also to have regular wheel alignments to keep your car rolling nicely.

So how regular is regularly? Well, you don't need to get an alignment in any specific time period or base your alignment on how many miles you have driven. However, it's better to be safe than sorry.

Some mechanics recommend you to have your alignment checked at least once a year. Most shops (especially tire shops that specialize in tires and alignment) will do a quick test or alignment check for free. A lot of popular tire shops also have deals like "buy one alignment, and get free alignments for the full life of your car". Make sure you check these out and you won't have to pay for any more than just one alignment.

When your wheels aren't aligned properly, your tires will wear out sooner. This causes you to have poorer handling, and can also cause unnecessary wear to the rack and pinion and other steering components. Again, this can mean big costs in the future.

If you have a 4x4 or do a lot of off roading (or consistently drive on rough roads), you should have your wheel alignment checked more often (more than once per year is recommended).

And of course, if your vehicle ever starts pulling more to either the left or right, get your alignment checked before it causes even more damage to your car. Sometimes it's not the alignment at all, and can come down to uneven tire wear (which can be caused by all sorts of things). And as always, if you think you need your car checked out, you probably need it checked out.

Jim Gaines

46

Topping Off Your Brake Fluid

Some people will tell you that topping off your brake fluid is "stupid" or "dangerous". Wrong and wrong.

Although you don't need to top off your brake fluid, it is safe to. It is never safe to let your brake fluid run down to nothing, of course for the reason that your brakes could fail while you're driving!

Each month you should check your brake fluid. It only takes one minute per month. Base your findings off of the manufacturer's recommendations of course, and before opening the master cylinder lid, make sure you wipe away the dust and dirt around the area.

If you don't wipe away this dust, it could contaminate your brake fluid, and dirty brake fluid could mean problems with your brakes.

Never substitute fluids. Don't use power steering fluid in the place of brake fluid, that's one of the worst things you can do. Finally, never use brake fluid that has already been opened previously, because once exposed to air, it can again become contaminated quicker than you think.

47

Anti Lock Brake Care

Another thing you can do to take care of your car is caring for your anti lock brakes.

The anti lock brake system that is found in modern cars (90s+) is very sensitive to moisture. Water and moisture can easily destroy your expensive ABS pump and cause the inside of the brake lines to rot.

It's not your fault! Since brake fluid tends to attract moisture over the course of a couple years, your brake lines should be bled in that same time period.

Also, when you are getting your free wheel alignment (as you learned previously you can usually find), ask them to check your brakes as well (but make sure you ask if it will cost you - some charge, some don't).

And similar to wheel alignment, if you have a 4x4 or you spend time off roading or driving on rough roads, you should have your anti lock brakes checked more often.

Jim Gaines

48

The Easy Oil Change Guide

If you don't know how to do a complete oil change, now is the time to learn.

Use this quick guide to do an oil change on most cars, or just learn the process so you know what you are paying for and how it is done.

First, your vehicle owner's manual should give you a basic guide, as well as should show you where the drain plug and dip stick and other parts of your car are. Make sure you know what you are doing before you remove anything from your car.

Start by draining your old oil. Just remove the drain plug (it usually screws off), and let the oil drain into a bucket, container, or drain pan.

Next, clean the drain plug on the oil pan, and wash it off before you reinstall your oil pan.

Once the drain plug is reinstalled, add in the recommended oil to the recommended amount.

To check your oil, run your car for about 10 minutes so that the oil can warm up. Then, park your car (or make sure it is) on level ground.

Then, turn the engine off, and wait another 10 minutes or longer so the oil can drain back to the oil pan.

Remove the dipstick, wipe it clean, and reinsert the dipstick and push it all the way in.

Again, remove the dipstick and read/check the oil level. It should be somewhere between the hash marks. If you need more oil, add a little more according to your manufacturer's specifications.

Finally, before taking your newly oiled car for a test drive, make absolutely sure that your oil drain plug is on tight, and your dipstick and other caps are secured. The last thing you want is a leak when you go to pull away.

49

When You Must Change Your Oil

Besides knowing how to give your car an oil change, it's also important to know when you should change your oil.

Today's manufacturers are actually starting to recommend a longer period between oil changes. However, the fact remains that the more dirt and metal particles that are removed from your oil and your engine, the longer it will last and continue to purr.

Changing your oil more often is more costly, but it also extends your engine's (and your car's) life.

If you want to maximize your engine's life, be sure to refer to your owner's manual for severe intervals of oil changing. You should also change your oil more often if you drive in stop and go traffic regularly.

For years it was recommended that your oil be changed every 3,000 miles. Those numbers and intervals are increasing in number, meaning you don't have to do it that often. However, there's no harm in sticking to the old numbers, as you can assure there will be no unnecessary problems with your car or engine because of low or old oil.

50

Which Oil, and Changing Filter

Before moving on to other areas of your car, I wanted to include a few more tips about which oil to use, different types of oil, and changing your oil filter.

One type of oil is detergent oil. Almost all modern oils are detergent oils, which remove dirt, dust and soot from the internal parts of the engines. This is why your oil becomes black, it is actually cleaning your engine. But when your oil gets too black, that means the oil is saturated, meaning it can't hold the dirt and soot particles any longer (which is why it is good to change your oil frequently).

Let's talk about oil viscosity. The viscosity of your oil is determined using 2 numbers. The first viscosity number is when the oil is cold. Before the second viscosity number, you will see the letter "W". Contrary to popular belief, W stands for WINTER, and NOT WEIGHT. The second number (following W) tells you the viscosity of the oil when it is at operating temperature. The oil gets thicker as the number gets larger.

Besides your car, you can choose your oil partly based on the climate and weather. Your owner's manual will list what oils are acceptable. If you live in a hot climate, 10W30 is an okay alternative to 5W30. And if you are using 10W30, but then the weather becomes extremely cold in winter, it is okay to switch down to 5W30 for the colder seasons.

As for changing your oil filter - follow what the manufacturer recommends.

But this doesn't mean you can't use a more high quality filter from a different brand, it just means there are different sizes. There are plenty of after market filters like Pennzoil, Motorcraft, Valvoline and more - just match the type/size to your car, and you'll be fine.

BEWARE: Quick note, your manufacturer filter will usually be of higher quality than after market filters, but there are still higher quality filters you can find in the after market - you just need to be smart, and be prepared to pay a little more for higher quality.

For synthetic oil users, premium filters are usually bought and used. Again (as you

67 Need To Know Tips To Extend Your Car's Life

would guess), these are higher priced, but the benefits of using a premium filter means better use of your oil, and less hassle in the future.

Jim Gaines

51

Fuel Filter Changing

Before I talk about anything in this tip/chapter, just know that fuel filters and oil filters are NOT the same thing.

Oil filters (what you learned about before) basically cleans the lubricant which keeps your engine running smooth.

Fuel filters on the other hand cleans out all the junk and buildup inside the fuel lines (your fuel system). This is mainly to keep the injectors clean, but also keeps your other fuel system parts clean as well.

Recently, manufacturers have been known to tell us that we don't need to change our fuel filters as often as we used to.

However, I still change my fuel filter once a year, and I recommend you do the same. You see, once a fuel filter gets clogged, it will cause your engine to perform very poorly, and will clog your injectors (very bad). This also reduces your gas mileage, which also sucks (not only sucks in general, but also sucks money out of your wallet, and you don't even know it!).

If a fuel filter hasn't been changed in years, the gas tank will begin to corrode, and you will see those corroded particles in the fuel filter.

52

Improve Gas Mileage With Simple Trick

One simple trick/tip to improve your gas mileage is having a clean air filter.

Check your air filter every couple/few months, and when it is dirty, make sure you replace it. Air filters are very cheap, and very easy to change on your own (don't take your car to a mechanic and have them charge you to do this).

With carbureted (as opposed to fuel-injected) engines and vehicles, all you have to do is remove the big metal lid, you really can't miss it (and if you are missing it, check that manual!).

With fuel-injected cars, you will instead remove the rectangular box. Again, your manual will show you exactly where you can find this.

The reason a clean air filter can improve your gas mileage is becaues a smooth, clean flow of fresh air is necessary in order for you engine to run efficiently. When you increase the size of your air intake, you are essentially increasing the performance of your vehicle.

Jim Gaines

53

Healthy Transmission Guide

There's not much you can personally do to keep your transmission healthy, but there is ONE THING you can do for sure.

To keep your transmission healthy, it is very important to change your transmission fluid regularly/frequently.

For new cars, it is recommended that you change your tranny fluid after the first 4,000 to 5,000 miles.

After that (or for used cars), it is recommended that you change your tranny fluid every two years (or every 20,000 - 25,000 miles).

Have you ever had a transmission blow or fail on you? I have, a few times - not fun. Most of the time, the transmission is the most expensive part of your car. Let it bust on you, and it'll bust your wallet, big time. So make sure you check the transmission fluid, and change it from time to time for healthy car care.

54

NEVER Overfill Crankcase

It is very important to never overfill the crankcase with oil. The crankcase is the housing for the crankshaft, and is usually located underneath the cylinders in your engine.

When you overfill your crankcase, air bubbles will begin to form in the oil, which in turn makes oil pump not be able to function or perform properly.

As always, your question is "What will happen to my car if this happens?"

Well, plenty of your engine components will become stressed due to overheating of the engine. This is what makes cars "blow up" in a sense.

This can also cause fouled spark plugs. When this happens, the firing tip (where the spark plug sparks) collects residue and gunk, which makes the spark plug 'foul'. This can also cause one or more of your spark plugs to completely stop working, which again is never good for your engine.

Jim Gaines

55

PCV Valve - Positive Crankcase Ventilation

Remember your PCV valve. PCV stands for Positive Crankcase Ventilation, and is a part of the emissions system (usually only found in older cars).

This valve re-circulates partially burned gases from the engine crankcase to the combustion chamber (for you, this isn't that important, but it is nice to know how it works).

In order to prevent the build up of sludge, as well as harmful corrosion, it is recommended to change your PCV valve about ever 30,000 miles. For some this will come quick, and for others it might never happen (if you sell your car or don't use it much).

By changing your PCV valve and preventing the buildup of sludge and preventing corrosion, you will also improve your car's gas mileage, which is always on our minds.

56

Oil And Transmission Coolers

If you use your vehicle to tow any sort of trailer, you should have an oil cooler installed (and it's also recommended to install a transmission cooler).

These types of coolers cost very little, are easy to install, and will save you lots of money in major engine and transmission repairs.

Basically, oil and tranny coolers simply cool down the liquid and fluid that runs through your transmission and your engine.

If you drive your car like most people, oil and tranny coolers aren't that necessary. However, if you are towing trailers, towing cars, or causing any extra strain on your engine (and transmission), it is recommended to install these coolers.

And before, when I talk about "transmission and engine repairs", you should know that a repair can a lot of the time cost more than replacing the part entirely. This is because of the labor costs that go into replacing tiny parts in a large car part. So take care of your car now, and avoid the extra costs later.

Jim Gaines

57

Get Some New Spark Plugs!

Once again, we are going to talk about getting better gas mileage. Over the years, this has become one of the biggest concerns of most new or used car buyers.

New spark plugs can definitely increase your gas mileage.

Now, electronic ignitions and cars with on board computers have really gotten rid of the need for the 'regular tune up'. However, it is still important to change your spark plugs from time to time.

Many manufacturers recommend you to change your spark plugs every 30 - 40,000 miles.

Spark plugs exist for the sole purpose of sparking your gas to make it explode, which makes your car GO. Good spark plugs means your engine will perform at it's best, and you'll enjoy the improved gas mileage.

58

Checking Your Hoses

There are a handful of hoses under the hood of your car, and you should be checking them from time to time.

Over time, hoses become brittle and can break away. You do NOT want this to happen, because this can cause major leaks in your car, and any time there is a leak means your engine can blow at any time.

When the car is shut off and has cooled, squeeze the hoses.

If the hoses are very stiff, make a crunch sound, are soft or sticky, look collapsed in any area or section, or have bulges in them, it means the hose has become too weak and should be replaced.

The coolant hose is one of the most important, because it cools the engine and keeps it at the optimum heat (so that it doesn't OVERheat).

You should never ever drive your vehicle when the coolant hose (or any other hose) is damaged, because your engine will overheat, and you could end up with a very expensive repair bill (not cool!).

Jim Gaines

59

Belt Tension and Tensioner

True story, one of my family members was driving only for a couple minutes (just pulled away from the house and started driving along the flat street), and all of a sudden she heard a loud snap in front of her.

At first she thought she ran over a small branch, and continued driving. That driving only continued for a few more seconds until she pulled over (her steering became extremely tight almost to the point of not being able to drive the car). We later figured out it was the power steering belt, and if she were to drive any further with it snapped, it could have very well damaged her car much more.

Just like the hoses in your car, you should frequently check the tension of all of your belts - belts for your AC compressor, power steering pump, as well as water pump.

It's easy to check for tension. Find the center of the belt where the longest exposed part is located (where it isn't touching any pulleys or wheels), and press it in at that area.

If you are able to press the belt down half an inch to 1 inch (but no more than that), then the tension is fine. If it doesn't depress at all, that means it's either stiff or cracked, or something might be caught in the belt. And if it depresses past one inch, it is much too stretched, and you should make an adjustment or get it replaced (most belts aren't too expensive, but it's cheaper to maintain it rather than buying new ones all the time).

Besides checking the tension, make sure you check for fraying and cracks in your belts. If there are any cracks (especially if they are all in one place), that means you should replace the belt immediately.

60

Checking Your Timing Belt

Besides the other belts in your car you learned about, I wanted to put this type of belt in it's own chapter tip.

The timing belt is (in my opinion anyway) the most important belt in your car.

I say this because when a timing belt fails, it usually results in thousands of dollars of engine damage. In order to avoid huge engine repair bills like this, just check your timing belt every once in awhile.

And besides engine damage, some cars will just turn off and become non-operational. This means if your timing belt fails, you'll be stuck on the side of the road with a car that doesn't work (call the tow truck!).

You don't have to check your timing belt every time you drive your car. In the past, it was recommended to change the timing belt at 50,000 miles. These days, some car manuals recommend checking at 75,000 all the way to 100,000 miles (check your own manual).

Again, you can check this belt like you did your others before. Check for fraying and cracking, stiffness - and especially check if any part of the belt is ripping. If it rips, it fails, and when it fails, the engine becomes damaged and stops working.

Jim Gaines

61

Properly Clean Your Engine

Over the course of the tips in this book, you've heard it over and over again - THE ENGINE.

I'm guessing even before this book, you probably already realized that the engine is the most important part of the car. It's like the heart and the brain (sort of). What I mean is, without the engine, nothing else (besides some electric components) in your car will work until it is fixed.

And on the other side of things, all it takes is just one 'thing' to happen to another part of your car, and it can start to damage your engine.

So, it's always a good idea to run a complete engine clean every couple of years. Not only will cleaning your engine make it look nice, and not only will it keep the parts under your hood last longer, but also by removing all of the dirt and grime, you are able to spot potential future leaks easier and quicker.

NOTE: DO NOT just open your hood and blast water all over your engine! My buddy did this once, and it wasn't too pretty. He spent hours trying to figure out what he flooded, what got gunked up, what became too dirty, what seized. It was a nightmare.

There are important engine components under your hood (like electrical parts, distributor caps, and more depending on the make of your car) that you don't want to get wet. An easy way to avoid this is to use plastic bags and/or electrical tape to cover these areas (just make sure you remove them before trying to turn on your car again).

Instead of just blasting water all over the place, you can take a small rag, liquid dish soap, and scrub away any of the grease and grime around your engine parts. If you'd rather be more legit, you can buy a "grease cutting detergent" from an auto or parts store/shop, and use that to clean your under-hood parts instead.

Quick tip: I've known many fellow mechanics that use GASOLINE to clean all PARTS on their cars. I wouldn't go overboard, but it's good to understand that gasoline has it's own cleaning agents inside (like detergent), and if nothing else, you can use

67 Need To Know Tips To Extend Your Car's Life

some gas to clean parts of your car.

Just make sure once your engine parts are clean that you use a dry rag to remove any wetness from under your engine. Although this isn't a huge deal, at least it will save you from smelling lots of different burning smells (once your engine gets hot).

Jim Gaines

62

Air Conditioning In Winter?

Am I crazy? Are you crazy? If you think I've lost my mind completely, I haven't.

Yes, you should 100% turn on your AC at least a few times in the winter.

You see, in winter (especially if you live somewhere where it snows), most people use their heat every day and every night, yet their AC will never be turned on (for obvious reasons).

However, when this happens, your AC compressor can seize (and seizing is bad!). If you don't want to have to purchase a new AC compressor every time winter is over, then make sure you turn on that AC for a few minutes, every now and then (once or twice a month).

Besides having to buy a new AC compressor (if you leave the AC off for long periods), you might also need to buy new hoses. This is because when no refrigerant (AC in water form pretty much) is able to circle through your hoses, they will become hard and brittle (and could crack and snap), which again can cause damage to your engine.

So, turn on that AC when you haven't in awhile. You can turn it on and go back inside your house for a few minutes, then come back and crank the heat if you want. This saves your AC compressor, your hoses, and makes sure your engine stays healthy as well.

63

Maintaining Your Car Battery

For some reason, this one guy always used to come to me with the same car battery problem. It seemed like every other week, he was calling me telling me his battery wasn't working, and he couldn't start his car.

After cleaning and sometimes replacing his battery numerous times, I told him about how he could maintain his car battery. He now doesn't visit me as much haha.

Just like any other important part of your car, it's necessary to keep your battery in tip top shape. Maintaining your car battery is not difficult at all. Here's some tips that may help...

If there is dirt or dust around the battery, wipe it with a damp rag using a dish detergent.

Also, it's important to clean the battery terminals (or posts). FIRST remove the negative cable (black!), then the positive (RED). Take a few tablespoons of baking soda, add a bit of water, and dip the brass wire battery brush into the mix (gives it a thorough clean and a fresh new look).

After that, make sure you check for cracks in and on the battery itself, as well as bulging of the battery. If there are any bulges or cracks, you must replace your battery immediately. After checking, reinstall the battery cables, this time starting with positive (RED, THEN black).

Another side note: Some batteries have 'vent caps'. If yours does, remove it and check the electrolyte level. The vent cap should cover the top plates of the battery by at least 1/2 inch. DO NOT use tap water (tap water contains damaging minerals) - Instead, use distilled water.

Jim Gaines

64

Leaky Radiator?

Leaky radiators suck. If you have a radiator that is leaking, there's a few ways of fixing the leak.

Note: Although these fixes should get you by, if the leak is too big, you should always see a trusted mechanic about the problem.

BEFORE doing anything to the radiator, and before removing the radiator cap, make sure your engine has cooled, otherwise you could get a shot of burning hot water or steam right in your face.

Check to see where the leak is. If the leak is in the radiator itself, you can buy radiator sealant (either in powder form or liquid form), or cold weld epoxy, or an egg, or black pepper.

All of those choices will work if the leak is in the radiator. Just open the radiator cap, pour in the sealant (or epoxy, or egg, or black pepper with SOME water), and it will form and harden where the leak is coming from, and comes in contact with air to form a seal, which will at least temporarily stop the leak from happening for awhile.

Now of course, if you have welding equipment (and are experienced with welding), you can simply remove the radiator fromthe car and weld the leak shut yourself (easiest if you know what you are doing).

However, if the leak is not in the radiator, but in one of your hoses that attaches to the radiator, the before-mentioned tricks won't help much at all. I would advise you to just replace to the hose (much cheaper than buying a brand new radiator). OR, if you just need a temporary fix, rap the leaky hose with duct tape (again, only if it is the hose, NO METAL).

65

About Coolant Diluting

This tip can be used by anyone, but this is mainly for those that live in colder weather.

Did you know that your cooling system NEEDS to include both water and coolant antifreeze? Put simply, you do not use coolant undiluted.

Usually the mix is 50/50, 50% coolant and 50% water.

On the other side, you should never ever just use straight water in your radiator.

Want another funny story? This one is super short... My buddy called me, said his car wouldn't 'go'. I already previously knew what he had done, and I warned him about it. "Did you get that coolant I told you to get?" I asked. "No, why?" he asked back. "Well, now your radiator, and probably other components connected to the radiator, are frozen."

In other words, if you live in cold weather, do NOT just add water. Water alone freezes, but water with coolant antifreeze well... makes it NOT freeze (get it?).

Especially in cold weather (but really anytime), check your coolant antifreeze to make sure you have the coverage you need, and to make sure you don't have to replace radiators and hoses and other parts of your car later.

Jim Gaines

66

Coolant Flushes

Speaking of coolant diluting, it's important to flush your coolant system. How often? Some coolants recommend every 2 years, others ever 5 years. I'd suggest at least checking it after 2 years, and changing it anyway to avoid unnecessary repair in the future.

Basically, coolant loses it's strength and eventually becomes contaminated over the years. So, read your coolant label for directions, and flush out the old coolant and add some new coolant antifreeze as well as water.

And as you are probably already anticipating, here's what can happen if you don't do a flush regularly (or ever): Your radiator can become damaged, your heater core can become clogged, and the water pump and thermostat will eventually fail as well.

ALL of these things can lead to engine damage, which as you know is never good. Coolant cools the parts of your engine, coolant is important, so make sure you flush your coolant ever 2 to 5 years (the earlier the better is my recommendation).

67

NEVER EVER Mix Coolants

Yep, we are still talking about coolant. Why? Because it is important, and I've seen too many mistakes from honest, regular people that causes major damage to their vehicles.

One of the biggest mistakes is mixing your coolant.

Now, sometimes mixing different brands of coolant is okay, but I still wouldn't recommend it. If you use a specific type of coolant, make sure you use the same coolant in the future, whether it is to flush your system or topping it off.

What is a BIG NO NO is mixing the COLOR of your coolants.

There are many colors of coolant, however the most common I've seen are green, pink, and orange.

The reason you don't want to mix coolant colors is because they are entirely different substances, and have different chemicals inside each. If you mix pink with orange, orange with green, green with pink, and so on, your coolant will form into a gelly (gel-like) substance, and instead of liquid running through your system, the gel will stop coolant from circulating.

When coolant stops, so does other parts of your car. Parts seize, stop working, overheat, and it's the engine that takes the hit in the end. So, NEVER mix the colors of your coolant, and you (as well as your car) will be fine.

Jim Gaines

Thank You For Reading!

Hey, just wanted to tell you "Thanks for reading", seriously. A few years ago when I thought about writing this book, it was hard because of the projects I was working on and I didn't seem to have the time to write.

It was when I was hearing more and more (seemed like every day) that my friends and family were having these car problems that I knew could be solved if they had just done one or two little things differently when taking car of their car. It was then I decided that this book needed to be written.

So, to my friends and family, and to you who just read this book - Thank you for reading it. I'm so happy I could help, even if it means doing the little things in order to prevent big problems from happening in the future.

Go on now, you have more knowledge than most have with cars. Share the knowledge, and let your friends and family know about these tips so they don't have unnecessary car problems in the future either.

www.ingramcontent.com/pod-product-compliance
Lightning Source LLC
Chambersburg PA
CBHW071756170526
45167CB00003B/1053